STEM K-5

SIMPLE MACHINES

wedges

VALERIE BODDEN

Published by Creative Education
P.O. Box 227, Mankato, Minnesota 56002
Creative Education is an imprint of The Creative Company
www.thecreativecompany.us

Design and production by Liddy Walseth
Art direction by Rita Marshall
Printed by Corporate Graphics in the United States of America

Photographs by Alamy (Art Directors & TRIP), Dreamstime (Ragne Kabanova),
Getty Images (David De Stefano, Mark Douet, Robert Harding, Johner,
Keystone, Steven Puetzer, Lauri Rotko, Clive Streeter, WIN-Initiative),
iStockphoto (Ivan Ivanov, Craig Uglinica, John Weise)

Library of Congress Cataloging-in-Publication Data
Bodden, Valerie.
Wedges / by Valerie Bodden.
p. cm. — (Simple machines)
Summary: A foundational look at wedges, explaining how these simple machines
work and describing some common examples, such as shovels, that have been
used throughout history.
Includes index.
ISBN 978-1-60818-012-7
1. Wedges—Juvenile literature. I. Title. II. Series.
TJ1201.W44B634 2011
620.8—dc22   2009048861
CPSIA: 040110 PO1140

First Edition
2 4 6 8 9 7 5 3 1

CREATIVE C EDUCATION

SIMPLE MACHINES

wedges

VALERIE BODDEN

contents

Have you ever dug dirt with a shovel or cut food with a knife? You might not have known it, but you were using a wedge. A wedge makes it easier to split objects.

A wedge is a kind of simple machine. Simple machines have only a few moving parts. Some have no moving parts at all. Simple machines help people do WORK.

An ax splitting wood acts as a wedge

A wedge is shaped like a triangle. It is made of two INCLINED PLANES joined together. The ends of the inclined planes come together to make a sharp tip, or blade.

A knife with a large blade can cut thick bread

To use a wedge, a person pushes or hits it into an object. As the person pushes down on the wedge, the blade forces the object to separate, or split. The object splits even more as the thicker end of the wedge is pushed in.

A long, thin wedge is easy to push into an object. But it may take more cuts to split the object. A short, wide wedge is harder to push into an object. But it may split the object faster.

Long, thin nails can be used as wedges

Wedges can be used to hold things in place, too. When a wedge is pressed tightly against an object, it keeps the object from moving. A wedge placed between a door and the floor holds the door in place.

Many doorstops are made of wood

People have been using wedges for thousands of years. Some of the first wedges were used to split wood. People put wooden wedges into cracks in rocks and soaked the wedges with water. As the wedges SWELLED, they split the rocks apart.

Today, metal wedges are used to split rocks into pieces

16

People still use wedges today.

The blades of scissors are wedges.

So are the blades of knives. Even

fork TINES are wedges!

An ax is a wedge. The tip of a nail is a wedge, too. Wedges are everywhere. Without them, it would be much harder to split the objects around us!

A CLOSER LOOK at *Wedges*

THE NEXT TIME YOU ARE EATING, TAKE A MOMENT TO LEARN MORE ABOUT WEDGES (WITH A GROWN-UP'S PERMISSION). FIRST, TRY TO STAB A PIECE OF FOOD WITH THE HANDLE OF YOUR FORK. DOES IT WORK? NOW STAB THE FOOD USING THE FORK'S TINES. WHY DO THE TINES GO INTO THE FOOD WHEN THE HANDLE DOESN'T? LOOK CLOSELY AT THE SHAPE OF THE TINES FOR YOUR ANSWER!

Glossary

inclined planes—simple machines made up of flat surfaces that are higher at one end than the other

swelled—got bigger; wood swells as it soaks up water

tines—the prongs, or thin, pointed parts of a fork

work—using force (a push or pull) to move an object

Read More

Oxlade, Chris. *Ramps and Wedges*. Chicago: Heinemann Library, 2003.

Thales, Sharon. *Wedges to the Rescue*. Mankato, Minn.: Capstone Press, 2007.

Web Sites

MIKIDS.com

http://www.mikids.com/Smachines.htm

Learn about the six kinds of simple machines and see examples of each one.

Simple Machines

http://staff.harrisonburg.k12.va.us/~mwampole/1-resources/simple-machines/index.html

Try to figure out which common objects are simple machines.

Index